CUPM Panel on Teacher Training

RECOMMENDATIONS

ON THE

MATHEMATICAL PREPARATION OF TEACHERS

The Mathematical Association of America

Committee on the Undergraduate Program in Mathematics

© 1983 by

The Mathematical Association of America
Library of Congress Catalog Card Number 83-062079
ISBN Number 0-88385-052-4

Printed in the United States of America

CUPM Panel on Teaching Training

Henry L. Alder, University of California, Davis, CA
Joseph Cicero, Coastal Carolina College, Conway, SC
John Dossey, Illinois State University, Normal, IL
Marjorie M. Enneking, Portland State University, Portland, OR
Richard K. Guy, University of Calgary, Calgary, Alberta, Canada
Donald W. Hight, Pittsburg State University, Pittsburg, KS
Katherine P. Layton, Beverly Hills High School, Beverly Hills, CA
Bruce E. Meserve, University of Vermont, Burlington, VT (Chair)
Gail S. Young, University of Wyoming, Laramie, WY
Don Bushaw, Washington State University, Pullman, WA (ex-officio)

ACKNOWLEDGEMENTS

Specific thanks need to go to Professor Kenneth Joy, University of California, Davis, CA for his assistance with the computer course. An additional thanks needs to be given Ms. Dianne Brewer and the Mathematics Department at Illinois State University. Ms. Brewer typed, duplicated, and distributed an uncountable number of drafts of this document. Her time and the associated costs were all underwritten by the Illinois State University Mathematics Department.

As these recommendations are being written, there is a critical shortage of adequately qualified mathematics teachers, a shortage likely to last for years. One cause for the shortage is the small number of students completing mathematically sound programs of teacher preparation; another is the commercial and industrial demand for those competent in the mathematical sciences.

This shortage contributes to the unfortunate widespread practice, which is also a result of other conditions, of assigning teachers prepared in other fields to teach mathematics classes. The statistics on this phenomenon are alarming.

The Mathematical Association of America will be among those working for both short-range and long-range solutions to this very serious problem. The short-range solutions may necessarily involve compromises with the ideal. For example, for some time it may be necessary, because of the small numbers of students involved, to base many teacher preparation programs on standard undergraduate courses; and retreading programs for teachers not previously prepared to teach mathematics may cover considerably less ground than might be wished.

The authors of these recommendations nevertheless hope that, in spite of the present situation, a significant number of programs and teachers will be able to benefit from careful and direct consideration of the ideas presented here. The proposed curricula may also serve as a reminder of what is being compromised when compromise seems necessary.

RECOMMENDATIONS ON THE MATHEMATICAL PREPARATION OF TEACHERS

CONTENTS

Final Draft

INTRODUCTION

From its beginning, the Committee on the Undergraduate Program in Mathematics (CUPM) of the Mathematical Association of America (MAA) has been concerned with the mathematical preparation of elementary and secondary school teachers. Published recommendations from the CUPM Panel on Teacher Training have included:

Recommendations for the Training of Teachers of Mathematics (1961; revised, 1966);

Course Guides for the Training of Teachers of Elementary School Mathematics (1961; revised, 1968);

Course Guides for the Training of Teachers of Junior High School and High School Mathematics (1961);

Teacher Training Supplement to the Basic Library List (1965);

Recommendations on Course Content for the Training of Teachers of Mathematics (1971) [reprinted in A Compendium of CUPM Recommendations, MAA, Washington, n.d., vol. 1, pp. 158-202].

Other CUPM publications have dealt, at least in passing, with aspects of teacher preparation.

The recommendations on the mathematical preparation of teachers were widely studied and discussed, especially under an intensive program of regional conferences supported by grants from the National Science Foundation. [See, for example, Forty-one Conferences on the Training of Teachers of Elementary School Mathematics: A Summary, CUPM Report 15 (1966).]

The CUPM recommendations have played an important part in the recent evolution of teacher preparation in mathematics, but now, inevitably, show signs of age. Well aware of this, CUPM decided in 1979 to reactivate its Panel on Teacher Training (hereafter called "the Panel"), and the recommendations that follow are the principal result of numerous meetings, voluminous correspondence, and far-ranging consultation by the Panel over the past several years. Faithful to the spirit of previous CUPM recommendations, but tuned more finely to anticipated needs of the eighties, these recommendations are intended to provide a clear, reasonable, and up-to-date guide to the preparation of mathematically well-qualified teachers for the years ahead.

Among the trends and events indicating a need for new recommendations were the increased emphasis on applications of mathematics and problem solving, the evolving content of college mathematics courses, and the increased availability of computers, large and small. See, for example, the report written by Alan Schoenfeld for the MAA Committee on the Teaching of Undergraduate Mathematics, Problem Solving in the Mathematics Curriculum: A Report, Recommendations, and an Annotated Bibliography, MAA, Washington, 1983. The need for new recommendations is also evident from the publication by the National Council of Teachers of Mathematics (NCTM; 1906 Association Drive, Reston, VA 22091) of An Agenda for Action: Recommendations for School Mathematics of the 1980's (1980) and Guidelines for the Preparation of Teachers of Mathematics (1981). The present recommendations may be regarded as an effort of the mathematical community to interpret NCTM's Guidelines in relatively concrete curricular terms.

Teachers of mathematics must have mastered a certain body of mathematics, be aware of its increasing relevance in our technological age, and be able to stimulate students to understand and use mathematics and to appreciate the significance of mathematical concepts. Since prospective teachers must be concerned not only with mathematical ideas but with the communication of those ideas to students, the Panel has been attentive to aspects of courses that are of particular importance for teachers as well as to the broader fabric of mathematical ideas.

For each of the three levels of instruction considered, the recommendations provide both a guide to preservice preparation and a basis for programs of further professional development.

The clienteles for the three levels are:

Level I   Teachers of elementary school mathematics

Level II   Elementary school mathematics specialists, coordinators of elementary school mathematics programs, and teachers of middle school and junior high school mathematics.

Level III Teachers of high school mathematics

The recommendations are designed to identify aspects of mathematics that are of particular importance relative to the specialized needs of teachers. These aspects are described in terms of semester courses consisting of at least 40 class hours. Objectives are listed for each course and numbered to facilitate reference to them in discussions of the courses. The ordering of objectives has no significance relative to their importance. The Panel assumes that the organization of student programs by courses, course titles, and the structure of courses will vary from institution to institution. The Panel accepts responsibility for _identifying_ minimal needs of teachers. The responsibility for _meeting_ the needs of teachers falls upon the institutions and the instructors.

SUMMARY OF RECOMMENDATIONS

Level I

Various combinations of the following four courses are recommended:

Course 1:   Fundamental Mathematical Concepts I

Course 2:   Fundamental Mathematical Concepts II

Course 3:   Geometry for Elementary and Middle School Teachers

Course 4:   Algebra and Computing for Elementary and Middle School Teachers

Teachers of early childhood mathematics (nursery school-kindergarten) should complete at least Courses 1 and 3; teachers of grades 1 through 6 should complete at least Courses 1, 2, and 3. The preparation of teachers of middle-school mathematics and teachers preparing for specialization in elementary school mathematics should be at Level II and include Courses 1, 2, 3, and 4.

Level II

The preparation of teachers at Level II should include a minimum of nine courses:

The four Level I courses

An introduction to calculus

Four more courses, other than calculus, selected from the Level III list of courses.

Students who have previous thorough mastery of the contents of Courses 1, 2 and 3 might take condensed versions of these courses. The introduction to calculus might be obtained either from Calculus 1 or from a liberal arts or business calculus course. Mathematics Appreciation, Discrete Mathematics, Probability and Statistics, Number Theory, and History of Mathematics are suggested as particularly appropriate electives.

Level III

The preparation of all teachers of high school mathematics should include at least the mathematical content of the following thirteen courses:

Discrete Mathematics

Calculus Sequence (three courses)

Introduction to Computing

Mathematics Appreciation

Linear Algebra

Probability and Statistics

Number Theory

Geometry

Abstract Algebra

History of Mathematics

Mathematical Modeling and Applications

Teachers of calculus should have additional work in analysis including the background for teaching the content of the Advanced Placement calculus courses.

All mathematics courses should be structured and taught so as to develop and maintain student interest and enthusiasm for mathematics. As prospective teachers the students in these courses must learn to communicate their knowledge of mathematics. Thus the courses should be designed to encourage active student participation. It may, for example, be appropriate to offer one or more of the courses in a seminar format in which students present selected material to the class. A problem seminar or student colloquium series could also provide opportunities for students to sharpen their skills in discovering, and communicating, mathematical concepts.

This 1983 Level III recommendation of the content of thirteen courses may be compared with the 1961 recommendation of eleven courses (nine specified and two selected from five electives) and the 1971 recommendations of twelve courses with selections suggested from an additional seven electives.

USES OF THE RECOMMENDATIONS

The components of the recommendations are identified in the descriptions of courses. Most of the courses may be found among the usual offerings for undergraduate majors in mathematics. The course descriptions in these recommendations are not intended to be complete course descriptions but rather to identify topics that are particularly important for teachers. The Panel encourages each institution to review its programs for prospective teachers and to identify in these programs the components of the recommendations. Since these are minimal recommendations, the Panel applauds those programs and institutions with more extensive requirements. In no case should these minimal requirements be used to justify restriction of more comprehensive programs.

Although these recommendations were developed for the prospective teacher, they may also be useful in designing inservice programs for teachers of mathematics who do not yet have this mathematical preparation. Such inservice programs may also serve elementary school teachers or secondary school teachers who are certified in other areas and wish to become mathematics teachers. Institutions are encouraged to make their programs available to such inservice teachers.

LEVEL I RECOMMENDATIONS

The past decade has brought changes to the teaching of elementary and middle school mathematics that have increased the number and variety of competencies demanded of teachers at these levels. The teacher needs better skills in the teaching of problem solving, estimation, concept development, and applications. At the same time, the need for preparation to teach the arithmetic of basic facts and algorithms has not been decreased. Thus, the modifications and additions to the K-8 curricula necessitate commensurate changes in the mathematical preparation of teachers for these levels.

In order that teachers may become adequately prepared to deal with the mathematical content at the levels that they are certified to teach, it is recommended that all prospective elementary school teachers complete the sequence of Courses 1, 2, and 3. Pre-service teachers completing a certification program in early childhood education designed specifically for pre-school aged children are encouraged to complete Courses 1 and 3. If their certification includes any of the grades in the K-8 range, they should also complete Course 2. Course 4, which might not ordinarily be expected of all prospective elementary school teachers of mathematics, is strongly recommended for pre-service teachers preparing for a specialization in elementary school mathematics and for students studying to teach middle school or junior high school mathematics.

The Level I sequence of courses assumes the prerequisite of three years of college preparatory mathematics consisting of two years of algebra and one year of geometry. It is desirable that prospective teachers choose additional Level II courses in order to prepare for specialization in elementary school mathematics. All students preparing for elementary school teaching are expected to complete at least one course in the methods of teaching mathematics at the elementary school level. While these recommendations do not detail the content of such a methods course, the NCTM Guidelines For The Preparation of Teachers of Mathematics describe the essential content of such a course.

The following four courses address various facets of a prospective elementary school teacher's mathematical needs. The content and its arrangement within the courses reflect current curricular patterns at colleges and universities, results of surveys of teachers' needs, and the content of current and evolving programs in elementary school mathematics.

Each course emphasizes the central role of problem solving in mathematics. Each course should cover at least the indicated topics with consideration of the approaches and objectives suggested. Problem solving should be highlighted through appeal to a variety of problem solving strategies in the teaching of the courses. Students should be provided with ample opportunities for explorations involving manipulative materials; analysis of nontraditional situations; and applications of central concepts, skills, and principles. Special attention should be given to the development of student competencies in estimation and mathematical judgment.

LEVEL I COURSE DESCRIPTIONS

Course 1: Fundamental Mathematical Concepts I

This course provides prospective teachers with part of the background
needed for teaching the content of contemporary elementary mathematics
programs including the development of the whole number system, geometry, and
measurement. The prerequisites are three years of college preparatory
mathematics, including two years of algebra and one year of geometry, and
demonstrated mastery of the basic algorithmic skills of the K-8 mathematics
curriculum.

Objectives

The objectives of Fundamental Mathematical Concepts I should be to
provide teachers with the ability to:

1. Identify, develop, and solve problems that are related to the
    child's environment and involve the mathematical concepts and
    principles usually taught in the elementary grades;

2. Identify and use problem solving strategies appropriate to these
    grades;

3. Illustrate prenumeration concepts (attributes, classification,
    ordering, patterns, and sets);

4. Illustrate and explain number and numeration concepts (cardinal and
    ordinal numbers, place value);

5. Explain and develop the usual algorithms for the four basic operations with whole numbers and illustrate these operations using models and thinking strategies appropriate to the elementary grades;

6. Relate the properties of the whole number system to the basic algorithms and to their use in problem solving;

7. Use estimation in numerical calculation and problem solving situations;

8. Identify examples in the child's environment of simple geometric shapes and their properties;

9. Develop basic planar relationships (parallelism, perpendicularity,...) and model them with examples from the child's environment;

10. Model spatial relations and illustrate their properties using classroom objects;

11. Use standard and nonstandard (paperclips, erasers, body measures, ...) units in measuring length, perimeter, area, capacity, volume, mass, weight, angle, time, and temperature;

12. Design classroom experiences illustrating the geometric and measurement concepts appropriate to the various grade levels;

13. Use mathematical terminology and symbolism appropriately while working with elementary school children;

14. Describe the historical and cultural development of some of the main mathematical concepts and principles usually taught in the elementary grades.

Topics

While there may be considerable variation in the topics selected to achieve these objectives, the following are particularly suitable for inclusion in such a course:

. PROBLEM SOLVING (4 hours)

Problem solving strategies (including guess and test, pattern searches, models, related problems, formulas, algorithms, and simulations), applications of strategies in traditional and nontraditional settings, preparation for use of strategies throughout the remainder of the course and following courses.

. PRE-NUMBER CONCEPTS (3 hours)

Properties of joining, separating, and comparing sets; set equivalence; set inclusion; Cartesian products; and the use of set concepts in problem solving (Venn diagrams, exhaustive listings,...).

. WHOLE NUMBERS AND THEIR OPERATIONS (15 hours)

Historical role of number systems, base ten numeration system, place value and its relation to grouping in operations, models for each of the four basic operations, properties of the basic operations, common error patterns in student computation with basic algorithms.

. NUMBER THEORY (3 hours)

Divisibility, prime and composite numbers, sieve of Eratosthenes, infinitude of primes, divisibility rules for whole numbers through 11 (excluding 7), lcm, gcd, and relative primeness.

. GEOMETRY (9 hours)

Basic concepts and properties of two- and three-dimensional spaces (including point, line, plane, space, segment, ray, betweeness, parallelism, perpendicularity, congruence, similarity, symmetry, and basic shapes and solids); use of geoboards, paper folding, and other models in illustrating these concepts and relationships; basic concepts and properties of geometric transformations (slides, flips, and turns).

. MEASUREMENT (6 hours)

Process of measurement (selection of a unit, covering with the unit, "counting" the number of units used); application of measuring using standard and nonstandard units of length, area, volume, capacity, mass, and their relationships; estimation of measures; perimeter, area, and volume of standard geometric figures; indirect measurement (similar figures, Pythagorean theorem,...).

Course 2: Fundamental Mathematical Concepts II

This course focuses on the development of the real number system and its subsystems, probability, statistics, and basic computer concepts. The prerequisite is Fundamental Mathematical Concepts I.

Objectives

The objectives of Fundamental Mathematical Concepts II are to provide teachers with the ability to:

1. Identify, develop, and solve problems that are related to the child's environment and involve the mathematical concepts and principles for the appropriate grade levels;

2. Explain the concepts of fractions (including decimals), integers, ratio, proportion, and percentage using models appropriate for the elementary grade levels;

3. Explain and develop the standard algorithms for the four basic operations for integers, positive and negative rational numbers (including decimal notation), and real numbers;

4. Illustrate the standard algorithms using properties of the number systems involved and appropriate models;

5. Recognize other algorithms for the basic operations and explain them using appropriate models or properties of the number systems;

6. Use estimation wherever appropriate, in particular to pose and select alternatives concerning reasonable responses;

7. Solve simple problems involving probability, inference, and the testing of hypotheses;

8. Solve simple problems involving the reporting of data (measures of central tendency, dispersion, expectation, and prediction);

9. Make appropriate use of calculators and computers in problem solving and in exploring and developing mathematical concepts;

10. Explain the role and the uses of computers in our society;

11. Use mathematical terms and symbolism appropriate to different elementary grade levels;

12. Describe the historical and cultural significance of some of the
major mathematical concepts and principles contained in the
elementary school mathematics curriculum.

Topics

While there may be considerable variation in the topics selected to
achieve these objectives, the following topics seem to be particularly
suitable for inclusion in such a course:

. THE REAL NUMBER SYSTEM AND ITS SUBSYSTEMS (14 hours--3 on integers, 5 on
rational numbers (fractions), and 6 on real numbers (decimals))

Extension of whole numbers to integers, models for integers,
operations with integers, properties of integers (order, absolute value,
...), extension of integers to the rational numbers, models for rational
numbers (fractional form), operations with rational numbers, properties
of rational numbers, decimal expansions of rational numbers,
irrationality of $\sqrt{2}$, irrational numbers, Pythagorean Theorem, extension
of the rational numbers to the real numbers, relation of types of real
numbers to forms of decimal expansions, properties of real numbers,
applications of real numbers, relationship of real numbers to classical
geometry problems, rounding and truncating real numbers on a calculator
and a computer, ratio, proportion, percentage, applications of
proportions in algebra and geometry.

. PROBABILITY (8 hours)

Relative frequency experiments, methods of counting (tree diagrams,
exhaustive listings, permutations, combinations), sample spaces, joint

events, independent events, dependent events, complementary events, Monte
Carlo methods.

. STATISTICS (6 hours)

Organization and presentation of data (line, circle, and bar graphs;
stem and leaf plots), role of scales and possible bias in graphs,
analysis of the observed distributions (mean, median, mode, range,
variance, standard deviation), box-and-whisker plots, sampling as it
applies to consumers of sample research literature.

. COMPUTER APPRECIATION (12 hours)

History of computing (Pascal to PASCAL), impact of computers on
contemporary school mathematics curricula, nature of an algorithm,
elementary programming (PRINT, REM, LET, INPUT, READ-DATA, IF-THEN,
FOR-NEXT, and GOTO), writing simple programs correlated to the elementary
school curriculum, role of computing in mathematics.

References for Fundamental Mathematical Concepts I and II

Bennett, A. B., & Nelson, T., Mathematics: An Informal Approach. Allyn
& Bacon, Boston, 1979.

Billstein, R., Libeskind, S., & Lott, J. W., A Problem Solving Approach
to Mathematics for Elementary School Teachers. Benjamin/Cummings, Menlo Park,
1981.

Meserve, B. E., & Sobel, M. A., Contemporary Mathematics. Prentice Hall,
Englewood Cliffs, 1981.

Shultz, J. E., Mathematics for Elementary School Teachers (2nd edition).
Merrill, Columbus, 1982.

Wheeler, R., Modern Mathematics: An Elementary Approach (4th edition).
Brooks/Cole, Monterey, 1977.

Course 3: Geometry for Elementary and Middle School Teachers

This course is designed to bridge the gap between secondary school mathematics and readiness to teach the geometry of the K-8 curriculum. As such, it is designed to build on the students' introduction to geometry in Course 1 by extending the topics discussed there and introducing additional topics. The prerequisite is Fundamental Mathematical Concepts I.

Objectives

The objectives of Geometry for Elementary and Middle School Teachers are to provide teachers with the ability to:

1. Identify, develop, and solve geometric problems that are related to the students' environment and involve the mathematical concepts and principles usually taught in the K-8 curriculum;

2. Recognize and construct consistent and logical arguments for geometric statements appropriate to the level of student being taught;

3. Solve simple problems in two- and three-dimensional geometry involving parallelism, perpendicularity, congruence, similarity, translation, reflection, rotation, symmetry, incidence, perimeter, area, and volume;

4. Construct simple planar figures and space models;

5. Present geometry as a source of models for real world phenomena;

6. Understand the relationship of geometry to other areas of mathematics;

7. Model the process of mathematical discovery through spatial examples (conjecture, testing, refinement, more testing, and final statement of result);

8. Extract geometric structures from situations having spatial relationships in them;

9. Explain the role of geometry in our cultural heritage--from both a scientific and an aesthetic viewpoint.

Topics

While there may be considerable variation in methods of reaching these objectives, the following topics would be suitable for inclusion in such a course, with at least one of the topics being covered in depth:

. GEOMETRY-ITS ROLES AND NATURE (4 hours)

Aesthetic and cultural aspects (weaving designs, Seurat's use of Golden Rectangle, architecture, ...), history of geometry, recreational geometry problems, types of geometry (Euclidean, projective, non-Euclidean and differential geometries, topology, ...), and unsolved geometry problems.

. EXPLORATION IN ELEMENTARY GEOMETRY (9 hours)

Major relationships and consequences, generating conjectures (through paper folding, geoboards, and other models), testing conjectures, constructions [straightedge and compass, plexiglass reflectors - (MIRAS)], properties of circles and related segments and lines.

. PATTERNS IN GEOMETRY (6 hours)

Numerical patterns in geometry (number of diagonals, angle sum in n-gon,...), tesselations (regular and semi-regular), finite differences and their relation to graphs of linear and quadratic functions.

. POLYHEDRA (5 hours)

Regular polyhedra, Platonic solids, tesselations in three-space, Euler's formula, geometric solids in nature (honeycombs, crystals,...).

. MOTIONS IN GEOMETRY (8 hours)

Translations (models, applications, and properties of translations, constructing images under a translation), rotations (models, applications, and properties of rotations, constructing images under a rotation), reflections (models, applications, and properties of a reflection, construction images under a reflection), composition of transformations, isometries and congruence, general classification theorem for isometries, generalization to three dimensions.

. SIMILARITY-MAGNIFICATION (3 hours)

Similar figures (plane and space), indirect measurement, magnification mappings, relationship of multiplier to size transformation, center of similarity, applications in projecting and pantographs, problem solving using similarity, similarity of three-dimensional figures.

. MEASUREMENT (5 hours)

Mensuration formulas (development and application); development of pi and its use in circumference and area problems with circle; estimation of geometric measurements.

References

Elliot, H. A., McLean, J.R. and Jorden, J.M., Geometry In The Classroom:
New Concepts and Methods. Holt, Rinehart and Winston of Canada, Toronto,
1968.

O'Daffer, P.G. and Clemens, S.R., Geometry: An Investigative Approach.
Addison-Wesley, Menlo Park, 1976.

Course 4: Algebra and Computing for Elementary and Middle School Teaching

This course is designed to bridge the gap between secondary school
algebra and the teaching of pre-algebra concepts in the intermediate and
middle school grades. In addition, this course extends the teacher's
knowledge of algebra through the use of computing to generate examples and
investigate relationships. The initial portions of the course are designed to
provide from an advanced standpoint a comprehensive view of the school algebra
curriculum. The remainder of the course is devoted to the investigation of
algebraic relationships and algorithms through the use of the computer. The
prerequisites are completion of the equivalent of four years of college
preparatory mathematics at the secondary school level, Fundamental
Mathematical Concepts I and II, and Geometry for Elementary and Middle School
Teachers.

Objectives

The objectives of this course are to provide teachers with the ability
to:

1. Develop the major algebraic properties and relationships of our number systems;

2. Explain the relationship between the algebraic and geometric views of major concepts in the K-8 curriculum;

3. Solve problems involving algebraic relationships found in the K-8 curriculum;

4. Engage in the development, analysis, and practice of algebraic skills in a variety of settings;

5. Master the major concepts related to functions and their graphs;

6. Master the major concepts and commands of programming in the language BASIC;

7. Apply the concepts and skills of programming in attacking algebraic problems;

8. Visualize algebraic relations through computer simulation and graphical output.

Topics

While there will be considerable variation in the topics selected to achieve these objectives and in the methods and materials used in teaching this course, the following topics would seem to be suitable for inclusion:

. FUNCTIONS AND CORRESPONDENCES (4 hours)

Concept of a mapping, functions (range, domain, graph), types of functions (1-to-1, onto, periodic,...), restricted domains.

. ALGEBRAIC FUNCTIONS (9 hours)

Linear functions (slope, y-intercept, x-intercept, general equations), quadratic functions (maximum, minimum, drawing tangents, orientation of graphs,...), methods of graphing (intercepts, relation of degree to critical points (described informally), behavior over an interval,...), roots.

. INTRODUCTION TO COMPUTING (3 hours)

Nature of an algorithm, relations of algorithms to computers, flowcharts, execution of flowcharts, types of input and output, blocks and structured programming.

. PROGRAMMING IN BASIC (15 hours)

Programming concepts and commands (statements, constants, variables--numeric & string, operations and relations, commands--PRINT, REM, LET, DEF, FOR/NEXT, READ/DATA, INPUT, GOSUB/RETURN; functions--SQR, RND, ABS, EXP, LOG, SIN, COS, TAN), graphics.

. APPLICATIONS OF COMPUTING IN ALGEBRA (9 hours)

Applications of computing to modeling and simulating algebraic situations (prime factorization of a whole number, solution of a linear equation, solution of a quadratic equation by various methods, solution of a system of linear equations, extraction of square root by Newton's method, area by Monte Carlo methods, geometric transformations), numerical analysis (rounding, truncation, amount of error, rational output rather than decimal,...), coding applications involving algebra and the computer.

References

Elgarten, G. H., Posamentier, A.S., Moresh, S.E. Using Computers in Mathematics. Addison - Wesley, Menlo Park, 1983.

LEVEL II RECOMMENDATIONS

Middle school and junior high school teachers must have the knowledge and skills needed to teach both elementary school mathematics and certain secondary mathematics courses, particularly algebra and geometry. They should also be prepared to teach special topics courses currently offered at some middle schools, such as computer literacy, beginning programming, elementary probability and statistics, studies of number patterns, and art and mathematics. To achieve these goals all middle school and junior high school teachers of mathematics need to complete at least the content of the following program:

A. Courses 1, 2, 3 and 4 for elementary teachers. (See the Level I section for a descriptions of these courses.)

B. An introduction to calculus.

C. Four courses (other than calculus) selected from the Level III list of courses. (See the Level III section for descriptions of these courses.)

A strong case could be made for requiring any one of the Level III courses. Mathematics Appreciation, Discrete Mathematics, Probability and Statistics, Number Theory, and History of Mathematics are suggested as particularly appropriate electives. At each college the selection of Level III courses for a prospective middle school or junior high school teacher should be based on the content of the available courses, their prerequisites and style of presentation, and the interests of the prospective teacher.

Students who have previous thorough mastery of the content of Courses 1, 2, and 3 might take condensed versions of these courses.

The introduction to calculus might be satisfied by completion of a portion of the Level III calculus sequence or by a survey of calculus intended for business, social science, life science, or arts and humanities majors. It should be noted that a few colleges have developed a special calculus course for middle school teachers. The most appropriate course is one that stresses the development of an intuitive understanding of limits, integration, and differentiation. Students should have the opportunity to use algebra and graphing skills in numerous simple examples and problems which illustrate the concepts; however, only the simplest techniques of integration and differentiation are recommended. The goal of the course is to develop an intuitive understanding of the powerful concepts of calculus and of their historical development. Throughout the course the students should encounter examples that help them relate the ideas to topics in middle school mathematics: approximating areas of irregular figures, simple related rate problems and the ideas of average versus instantaneous velocity, rational and irrational numbers and their decimal expansions, and the meaning of statements like "$4.\overline{9} = 5$."

LEVEL III RECOMMENDATIONS

The program of mathematics for the secondary school teacher as outlined in the following materials is based on the assumption that all students entering the program will have completed four years of college preparatory mathematics and should in particular have mastered a number of topics which have traditionally been taught as "College Algebra." Each institution implementing the following sequence of courses for prospective secondary school teachers should require that the students master the topics listed below either prior to or concurrently with the first course in the calculus.

These "College Algebra" topics permeate the fabric of undergraduate mathematics and a firm understanding of them is needed to teach secondary school mathematics. They are similar to the competencies for secondary mathematics students listed in the College Board document <u>Preferred Patterns of Preparation in Mathematics</u>, and are:

<u>Algebra</u>   Logical propositions, truth tables, patterns of inference; permutations, combinations, simple counting problems; the binomial theorem; trigonometric identities and equations; properties of, and operations with, complex numbers; trigonometric form of complex numbers and DeMoivre's Theorem; exponential and logarithmic equations; proof by mathematical induction; arithmetic and geometric sequences and series; matrices and determinants.

Geometry   Analytic geometry of the plane; transformations (translations and rotations); the conic sections; Law of Sines and Law of Cosines; vectors; polar coordinates.

Functions   Polynomial functions and their graphs; exponential and logarithmic functions and their graphs; circular and inverse circular functions and their graphs.

The content of the following courses constitutes the minimal recommendations for Level III preparation. Teachers of calculus should have additional work in analysis including the background for teaching the content of the Advanced Placement courses.

Discrete Mathematics

Calculus 1, 2, and 3

Introduction to Computing

Mathematics Appreciation

Linear Algebra

Probability and Statistics

Number Theory

Geometry

Abstract Algebra

History of Mathematics

Mathematical Modeling and Applications

The objectives of the courses are stated in the course descriptions and numbered to encourage discussion of them. A few relations among courses are stated in the course descriptions. The unity of mathematics should be emphasized at both the school and college levels. Far from giving the impression that it is natural and desirable to consider the subject of a particular course as something separate from the rest of mathematics, each instructor should emphasize ways in which ideas from other branches of mathematics are related to the subject under study. There are many potential relationships, in both directions, between courses. For example, should geometry precede algebra to provide examples of uses of matrices and groups or should algebra be studied first to provide tools of geometry? Should number theory precede algebra to provide examples of groups, rings, and fields, or should modern algebraic methods be used in studying number theory? No effort has been made to answer such questions in this report. In general an emphasis upon the application of mathematical tools in several types of situations is recommended. For example, something could be said about the use of matrices in relation to: systems of linear equations, rotations of axes in coordinate geometry, states and transitions in probability, linear programming, and two-person games.

LEVEL III COURSE DESCRIPTIONS

Discrete Mathematics

This course is a survey of topics in discrete mathematics. The prerequisite is described by the suggested high school preparation indicated in the Level III preface.

Objectives

The objectives of the discrete mathematics course are to enable prospective teachers to:

1. Develop basic techniques and modes of reasoning of combinatorial problem solving;

2. Describe and analyze the algebraic structures of certain set-relation systems;

3. Illustrate and analyze the wide variety of applications of discrete mathematics.

As described in the CUPM Recommendations for a General Mathematical Sciences Program (pp. 26-27) a discrete methods course "... should contain a variety of applications and use them both to motivate topics and to illustrate techniques." Computers, operations research, engineering, political science and transportation are some of the areas from which applications can be drawn. The spirit or style of this course is as important as the topics listed. To quote from the CUPM Recommendations again, "... the object of this course is not to show students simple answers. It is to teach students how to discover such simple answers (as well as not so simple answers). The means for achieving solutions are of more concern than the ends. Learning how to solve problems requires an interactive teaching style. It requires discussion of the logical faults in wrong analyses as much as presenting correct analyses."

Topics

The Panel recommends the following core of topics for a Discrete Mathematics course:

. MATHEMATICAL INDUCTION (3 hours)

Weak induction, strong induction, recursive notation.

. SET THEORY (4 hours)

Notation, basic properties, Venn diagrams, pigeonhole principle, truth tables, propositional calculus.

. RELATIONS AND FUNCTIONS (4 hours)

Binary relations, order relations, lattices, functions.

. COMBINATORICS (6 hours)

Counting arguments, permutations, combinations, principle of inclusion-exclusion.

. GRAPHS (7 hours)

Planar graphs, trees, paths and their use in searching, coloring, Eulerian & Hamiltonian graphs and their applications, graph based games.

. BOOLEAN ALGEBRA (4 hours)

Elementary definitions, examples and properties, principle of duality, lattices, switching circuits.

. DIFFERENCE EQUATIONS (5 hours)

Recurrence relations, generating functions, Fibonacci numbers and their applications.

In addition, it is suggested that at least one of the following topics be selected and treated with some depth:

. ELEMENTARY PROBABILITY

Basic laws of discrete probability, discrete probability distributions, random number generators, basic ideas of queueing theory.

. ALGORITHMS

Notation for expressing algorithms, analysis of algorithms.

References

Korfhage, Robert, Discrete Computational Structures. Academic Press, New York, 1974.

Lipschutz, Seymour, Discrete Mathematics. McGraw-Hill, New York, 1977.

Liu, C.L., Elements of Discrete Mathematics. McGraw-Hill, New York, 1977.

Preparata, Franco and Yeh, Robert, Introduction to Discrete Structures. Addison-Wesley, Reading, 1973.

Ralston, Anthony, Computer Science, Mathematics and the Undergraduate Curricula in Both. Technical Report Number 161, SUNY at Buffalo, July 24, 1980.

Stone, Harold, Discrete Mathematical Structures and their Applications. Science Research Associates, Chicago, 1973.

Tucker, Alan, Applied Combinatorics. J. Wiley & Sons, New York, 1980.

Calculus Sequence

Prospective teachers should take the same calculus courses taken by mathematics majors. They need to know the mathematics and applications of calculus and they need to develop the skills to work problems. Also, as prospective teachers, especially, they should appreciate and master topics from high school mathematics that are used in calculus, such as arithmetic, fractions, algebra, geometry, analytic geometry, elementary function theory, trigonometry, exponents and logarithms, curve sketching and calculator and computer utilization.

The developments of topics and levels of sophistication appropriate for teachers are described currently with phrases such as "intuitive understanding," "spiral approach," "developmental re-examination" as opposed to "rigor," "formal," and "theory." All teachers of mathematics (including the calculus instructor and high school teacher in training) are confronted with choices in developing topics at appropriate levels of sophistication. Prospective teachers should be informed in their study of calculus that choices in levels of rigor and methods of development are being made and that both intuition and rigorous proof have appropriate roles in mathematics.

Two important criteria in judging the appropriateness of developments of topics and levels of sophistication in the calculus courses are:

1. Mathematical correctness. The presentation should include only those statements that are mathematically valid and are stated in phraseology that is mathematically correct. An argument, however, may include assumptions or claims of existence that are only illustrated or made plausible by examples.

2.   Experience relatedness. Developments of topics should be based on students' previous experience and tied to related topics within their experience. Thus, a complicated concept such as limits can be presented by the "spiral approach," building greater understanding at successive encounters as students' knowledge and experience accrue.

The topics of calculus are typically taught in three courses. The sequence of topics typically includes the derivative (concepts, skills, and applications), the integral (concepts, skills, applications, and special techniques), series and parametric representation of functions, and multivariate calculus. The listing of topics below approximates the traditional order to illustrate one sequence. Development of a teachable re-ordering of topics is encouraged. Also, the breaks between courses could be moved forward effectively. The topics listed in courses below are minimal.

To illustrate the concepts of calculus and demonstrate its utility, current applications from other disciplines should be distributed throughout the sequence. Calculators should be incorporated into the study because they are part of students' experience. Computer programs should be used by students to facilitate finding answers and testing conjectures, and by instructors for illustrations, simulations and computations.

Calculus I

Calculus I is primarily a study of functions the incoming students know: polynomial, rational, algebraic, exponential, logarithmic (and trigonometric in either Calculus I or II). The derivative is a function, the antiderivative is a class of functions and the integral generates a function of one limit of integration. Functions are combined, analyzed, applied, graphed, inverted, differentiated, integrated and subjected to limits. The simple topic of functions unifies Calculus I nicely.

The prerequisites are two years of algebra including graphs of functions (especially exponential and logarithmic) and an introduction to matrices and determinants, and one year of geometry including traditional high school deductive geometry and coordinate geometry. A thorough study of trigonometry is required to precede the development of the derivative of sine and cosine functions and integration with trigonometric substitution.

Objectives

The objectives of Calculus I are to provide teachers with the competence to:

1. Master, by application and extensive use, the basic topics of high school mathematics, algebra, geometry and trigonometry;

2. Appreciate the role of real functions in society;

3. Use graphs to estimate related values, relative rates, extreme values, limits, derivatives and integrals;

4. Develop a concept of limits;

5. Understand the derivative, the integral and the Fundamental Theorem of Calculus;

6. Apply concepts and techniques of calculus to analyze functions and find relative rates, extreme values;

7. Use numerical methods to evaluate derivatives and integrals and use calculators and computers efficiently as tools;

8. Model problems from geometry and other disciplines using calculus notions.

Topics

Specific numbers of hours are not suggested for the topics in this and the two following course outlines. The Panel believes college mathematics instructors are so well acquainted with the teaching of calculus, and the context of calculus courses varies so much from one school to another, that a recommended allocation of time here would probably be neither necessary nor very useful.

. INTRODUCTION TO ANALYTIC GEOMETRY, FUNCTIONS, LIMITS, AND CONTINUITY

Cartesian coordinates; increments; slopes and linear equations; function, domain, range, graph, combinations as sums, differences, scalar multiples, products, quotients and composites, inverses; graph sketching techniques, symmetry, zeros, periodicity; limits and continuity with ϵ- and δ-notion based on graphing experiences and illustrated on calculator and computer; asymptotes; average velocity; secant line to a graph.

. THE DERIVATIVE AND ITS APPLICATIONS

Instantaneous velocity; tangent line to a graph; derivatives of polynomial functions and derivative of a scalar multiple of a function to emphasize the definition; derivatives of sums, differences, products, and quotients of two functions; the chain rule; rational and algebraic functions; increasing or decreasing functions; extrema; first derivative test; the derivative as a function; informal graphical presentations of cos x as the derivative of sin x and −sin x as the derivative of cos x; presentations of derivatives of other trigonometric functions as quotients or reciprocals; continuity-differentiability relationship; implicit differentiation; higher derivatives; concavity; points of inflection; acceleration; second derivative test; illustrations of

the intermediate value theorem and of the existence of maxima or minima;
proofs of Rolle's theorem and Mean Value Theorem; Newton's method.

. ANTIDERIVATIVES

Definition and simple examples; velocity from acceleration, and
position from velocity, with initial conditions.

. LOGARITHMIC AND EXPONENTIAL FUNCTIONS

The natural logarithm as antiderivative of x^{-1} such that ln 1 = 0;
e; exponential functions; logarithmic differentiation; exponential growth
and decay; the differential equation $y' = ky$.

. THE DEFINITE INTEGRAL

Riemann sums; definite integral; area; properties of the definite
integral; Fundamental Theorem of Calculus with historical comments; chain
rule, mean value for integrals; numerical integration with trapezoids;
areas between curves; volumes by disks, shells and crossections.

Calculus II

Calculus II is primarily an extension of the calculus the students
learned in Calculus I. Greater mastery of methods of integration is gained by
various techniques of integration and by numerical methods. Inverse functions
are emphasized and surprising relationships between algebraic and inverse
trigonometric functions are learned. Calculus is used to represent functions
with series which are then studied using calculus and numerical methods.
Especially, the basic notions of derivative and Riemann integral are applied
to model situations in geometry, physics and other disciplines and to extend
calculus to polar coordinates and parametric representations. For Calculus II

to be a unified study rather than a grab bag of topics, emphasis must be placed on the calculus notions being applied and on the results of numerical methods, rather than on the techniques or computer programs which generate values or the variety of topics being modeled.

The prerequisite is Calculus I.

Objectives

The objectives of Calculus II are to provide teachers with the competence to:

1. Master, by application and extensive use, the basic topics of Calculus I;

2. Develop derivatives and antiderivatives of trigonometric and hyperbolic functions;

3. Develop derivatives of inverse functions and apply formulas to integrals;

4. Evaluate integrals by applying tables, numerical methods and techniques of parts, partial fractions and trigonometric substitution;

5. Test sequences and series for convergence;

6. Express functions with series and then differentiate, integrate and evaluate the functions;

7. Apply limit notions to error estimations and apply derivatives to evaluating limits;

8. Extend Calculus I notions to parametric representations, polar coordinates;

9. Model problems from geometry and other disciplines with calculus notions.

Topics

. TRIGONOMETRIC FUNCTIONS

Limits of trigonometric functions, (sin x)/x, derivation of derivatives, antiderivatives, inverse trigonometric functions, rotation of axes, parametric representation, polar coordinate representation.

. TECHNIQUES OF INTEGRATION

Techniques of integration; change of variable (from the derivative of a composite); parts (from derivative of a product); partial fractions; trigonometric substitution with emphasis on inverse functions; use of tables; numerical methods of integration.

. APPLICATIONS OF THE DEFINITE INTEGRAL

Application of the Riemann integral to model areas, volumes, centroids, etc.; improper integrals, L'Hospital's theorem; comparison of rates of convergence.

. INFINITE SERIES

Taylor's formula; error analysis; sequences; limits; computer illustrations with general and recursion formulas; Newton's method; series of real numbers; geometric series; integral, comparison and ratio tests; alternating series; absolute and conditional convergence; approximations and errors; derivations of Taylor's series; functions defined by series; substitutions, derivatives, integrals and approximations with polynomials.

. POLAR COORDINATES

Polar coordinates; parametric equations; curves; tangent lines; conics; Riemann integral models in polar coordinate systems.

Calculus III

Calculus III consists of typical multivariate topics. For teachers, one primary purpose is the development of spatial concepts and 3-dimensional graphing techniques. The unity of multivariate calculus is in the recognition of spaces under consideration. This point should not be missed by teachers or students. Functions from the plane (R^2) to the real numbers (R) generate surfaces in 3-space and are studied with vectors in R^2 and R^3, partial derivatives, total derivatives, tangent planes, volumes and directional derivatives. The mathematics is easily extended to functions from R^n to R. Functions from R to R^2 or R to R^3 generate paths in two- or three-space. The study involves parametric representations, vector fields with various applications to velocity, force, etc., line integrals, tangential vectors, divergence and curl. Within the overall theme of multivariate systems, topics arise naturally (vectors, coordinate systems, surfaces and lines in space, tangential planes and vectors), extensions to higher spaces are anticipated (R^n to R^m, use of linear algebra, determinants, Jacobian, Hessian) and Green's theorem evolves as an exciting, surprising relationship.

The prerequisite is Calculus II.

Objectives

The objectives of Calculus III are to provide teachers with the competence to:

1. Master, by application and extensive use, the basic topics of Calculus I and II;

applications; vector fields, line integral paths.

References

Further discussions may be found for the first and second courses respectively in

Advanced Placement AB Calculus Syllabus,

Advanced Placement BC Calculus Syllabus, Educational Testing Service, Princeton,

and for all three courses in the CUPM report:

Recommendations for a General Mathematical Sciences Program, The Mathematical Association of America, Washington, 1981.

Introduction to Computing

Secondary school teachers should view the use of computers as a standard mathematical procedure with significant applications in nearly all aspects of society. The major emphases in this course should be the design and implementation of several representative computer applications through the vehicle of a high-level computer language (PASCAL). The prerequisite is four years of college preparatory mathematics including the topics listed in the Introduction to Level III Recommendations.

Objectives

The objectives of the introductory computing course are to provide teachers with the ability to:

1. Design and implement computer programs in a variety of areas of mathematical applications;

2. Use fluently a high level structured programming language (PASCAL);

3. Explain the computer's roles, and related methods, in applying mathematics.

The following four major topics provide a map for the content of the course. Course instruction should be in the context of applications with an intermixing of aspects of the topics of applications, language specifics, and design. If the instructor thinks that it is necessary to give a segment on number systems or on the internal machine language, these should be put off until the language specifics and a substantial part of the design and control have been finished. Then students can relate to a model of a machine (the PASCAL language) and the instructor can assign some programs that simulate the activities of a machine.

Topics

While there may be considerable variation in the topics selected to achieve the above plan of study, the following four areas seem particularly suitable for inclusion:

. INTRODUCTION (6 hours)

Hardware (central processor; memory—primary and secondary; peripherals—input, output, and special, i.e., graphics); languages available; operating systems overview (editors, compilers, linkers, filers and file concepts).

. LANGUAGE SPECIFICS (12 hours)

Data types; expressions (algebraic and Boolean (logical)); control structures; input and output (interactive and file manipulation); data structures (arrays and strings); procedures and functions; documentation.

. DESIGN AND CONTROL (8 hours)

Logical organization; design considerations (top-down, stepwise refinement); control (testing and test plans, bottom-up testing, verification, debugging techniques, documentation, structure considerations, style considerations).

. APPLICATIONS (18 hours)

Algebraic examples (prime numbers, perfect numbers, etc.); user-oriented, interactive systems; statistical data analysis; probability analysis through Monte Carlo methods; simulation; finding zeros of a function; sorting and searching an array; elementary graphing; elementary text processing; file processing.

References

Braineed, Walter S., Goldberg, Charles H. and Gross, Jonathan. PASCAL Programming - A Spiral Approach. Boyd and Fraser Publishing Company, San Francisco, 1982.

Clark, Randy and Koehler, Stephen. The UCSD PASCAL Handbook. Prentice-Hall, Englewood Cliffs, 1982.

Schneider, G.M. and Bruell S.C. Advanced Programming and Problem Solving with PASCAL. John Wiley and Sons, New York, 1981.

Tremblay, Jean-Paul and Bunt, Richard F. An Introduction to Computer Science: An Algorithmic Approach. McGraw-Hill, New York, 1979.

Trembley, Jean-Paul, Bunt, Richard F. and Opseth, Lyle M. Structured PASCAL. McGraw-Hill, New York, 1980.

Mathematics Appreciation

This is a course to introduce prospective teachers to a broader range of mathematical topics than is ordinarily presented in the more technical courses of the standard undergraduate mathematics curriculum. Topics are selected from many branches of mathematics. The prerequisites are two years of high school algebra and one year of high school geometry.

Mathematics appreciation courses are primarily offered for students who have a minimal mathematical background. There are many students with a great variety of majors for whom a mathematics appreciation course should be available. Such a course, taken during the freshman or sophomore year, would be most beneficial to the prospective teacher. By presenting a variety of mathematical topics, it can offer students valuable guidance in the choice of advanced electives and provide some additional preparation for such courses.

For prospective teachers in institutions where such a course is not available, a junior or senior seminar could be an appropriate alternative to a mathematics appreciation course. Such a seminar could provide students with the opportunities to explore a variety of topics not normally encountered in the standard undergraduate curriculum or to investigate standard topics in greater depth. It could serve as an opportunity for students to develop expository skills by preparing such topics for class presentation. Such a seminar is valuable also for students after they have taken an elementary mathematics appreciation course.

Either a mathematics appreciation course or a seminar can serve for the prospective teacher as an excellent model of how to present interesting and exciting topics in mathematics to high school students to enrich their classroom experiences.

Objectives

The objectives of the mathematics appreciation course are to provide teachers with:

1. An appreciation of the significant role that mathematics plays in society, both past and present;

2. An understanding of some exciting parts of mathematics which they are not likely to encounter in the typical curriculum of a mathematics major;

3. A knowledge of some topics of mathematics that require little mathematical background and can be used in school courses as devices to lend enrichment and excitement to their own teaching.

Topics

Since mathematics appreciation courses are taken by a large number of students, the CUPM established in 1977 the CUPM Panel on Mathematics Appreciation Courses to consider the content of these courses. The report of that Panel was approved by CUPM in 1981. The reader is referred to this report (See American Mathematical Monthly, vol. 90, 44 – 51, 1983) for a discussion of the philosophy of such courses, things to stress, things to avoid, approaches to course organization, examples of topics, references, etc.

While the Panel on Mathematics Appreciation Courses has many recommendations to offer, it does not hold that a particular selection of topics or teaching strategies should be universally adopted for mathematics appreciation courses. All material presented in such courses should be well motivated and related to the role of mathematics in culture and technology.

Standard textbooks offer a rather traditional assortment of topics: probability, graph theory, number theory, finite difference equations, topology, computers, matrices, statistics, exponential growth, set theory, and logic seem to dominate. But there are numerous other themes that can be used for large or small components of courses. Here are a few of the many possible examples. Seven of these are taken from the CUPM Report of the Panel on Mathematics Appreciation Courses. For each topic at most 6 hours should normally be allotted, for some considerably less.

. UNDERSTANDING HOW TO USE THE BUTTONS ON A POCKET CALCULATOR

It used to be that the number e was a complete mystery to those who had not studied calculus, and that "sin" had for humanities students more the connotation of theology than of mathematics. But no more. Virtually everyone has, or has seen, inexpensive hand calculators with buttons that perform operations involving exponential, trigonometric, and basic statistical functions. Teaching a class what these buttons do is an exciting new way to explore some traditional parts of classical mathematics.

. TRACING THE MODERN DESCENDENTS OF CLASSICAL MATHEMATICAL IDEAS

This can illustrate both the power of mathematics to influence the real world and the remoteness of mathematics from the real world. For example, classical Greek geometry involving conic sections led to models for planetary motion, and ultimately to the possibility of space flight. And probability, which had its origins in seventeenth century discussions about gambling, now dominates actuarial and fiscal policy, influencing government and corporate budgets, thus affecting the level of interest, the level of unemployment, and the health of the entire economy.

. CONNECTING MATHEMATICS WITH NOBEL PRIZES

Nobel prizes are not given in mathematics (and the apocryphal reasons for this are quite amusing). But the work that led to Nobel prizes (e.g., of Libby, of Arrow, of Lederberg, and others--see the articles in Science on Nobel prize winners) often has an intrinsically mathematical basis. The study of this scientific work provides an opportunity to show how mathematics is important in the most profound discoveries of modern science.

. APPLYING EXPONENTIAL GROWTH MODELS

The applications of traditional topics from elementary mathematics can often be explored more fully than has usually been the case. Exponential growth and decay models provide a striking example. Simple non-calculus approaches to models of growth provide a basis for discussion not only of interest and inflation, but also of such things as radiocarbon dating, cooling and heating of houses, population dynamics, strategies for controlling epidemics, and even detection of art forgeries.

. RELATING TRADITIONAL MATHEMATICS TO NEW APPLICATIONS

A discussion of beginning probability theory can quickly lead to a treatment of the Hardy-Weinberg law of genetics and a calculation of the probability of winning state lotteries. An introductory treatment of statistics can quickly lead to a discussion of political polls, of the design and interpretation of surveys, and of problems in decision theory. Modern applications of elementary network theory include recent work in computational complexity and "almost unbreakable" codes.

. INTRODUCING PROBLEMS INVOLVING DECISION-MAKING

There are many situations described by elementary mathematics in which one must choose "rationally" among possible options. One can discuss quantifying risk and uncertainty, fair division schemes, applications of network flows, pursuit and navigation problems, game theory and numerous other topics. Political science is full of unexpected but usually interesting topics, including Arrow's theorem and its offshoot theories of voting, the recently discovered problems associated with apportionment of legislatures, and strategies of fair voting in multiple candidate elections.

. EXPLORING THE POWERS AND LIMITATIONS OF MATHEMATICAL MODELS

Each of the modern social sciences abounds with applications of elementary mathematics. All of the examples mentioned above, and many more, involve the use of mathematical models. Sometimes these models are quite accurate and sometimes they are not. But even in the latter case the model can help clarify one's thinking about the underlying problem. An example of this use of mathematical modeling is the prisoner's dilemma argument of game theory and its possible connection with U.S. - U.S.S.R. relations.

. MAP COLORING

Coloring maps with two colors, proof of the five color theorem, discussion of the four color theorem.

. INFINITE SETS

Sets and one-to-one correspondence, finite and infinite sets, natural numbers as cardinalities of finite sets, very large natural numbers (grains of sand, drops of water in a lake, etc.), denumerability of the rational numbers, Cantor's proof of the nondenumerability of the real numbers, comments on recent discoveries concerning transfinite numbers.

References

For an extensive list of films (together with their distributors), classroom aids, and survey monographs, essays and other references, see the report of the CUPM Panel on Mathematics Appreciation Courses.

There is probably no single textbook covering all the topics suggested above, but most of them are covered in at least one of the following texts or books:

Courant, Richard and Robbins, Herbert. What Is Mathematics? Oxford University Press, London, 1941.

Davis, Philip J. and Hersh, Ruben, The Mathematical Experience. Birkhauser, Boston, 1981.

Messick, David., ed., Mathematical Thinking in Behavorial Sciences. W. H. Freeman and Company, San Francisco, 1968.

National Research Committee on Support of Research in the Mathematical Sciences (COSRIMS). The Mathematical Sciences – A Collection of Essays. MIT Press, Cambridge, 1969.

Rademacher, Hans and Toeplitz, Otto, The Enjoyment of Mathematics. Princeton University Press, Princeton, 1957.

Schaaf, William L., ed., Our Mathematical Heritage (New Revised Edition). Collier Books, New York, 1963.

Steen, Lynn Arthur, ed., Mathematics Today: Twelve Informal Essays. Springer-Verlag, New York, 1978.

Stein, Sherman K., Mathematics, The Man-Made Universe: An Introduction to the Spirit of Mathematics (3rd Edition). W. H. Freeman and Company, San Francisco, 1976.

Linear Algebra

This is a study of matrices, vector spaces, linear transformations, and numerical aspects and applications of them. The prerequisite is two semesters of calculus. (If three semesters of calculus are required, emphasis should be shifted to the more advanced topics.)

Objectives

The objectives of the course are to enable prospective teachers to:

1. Understand the basic theory of finite dimensional vector spaces and linear transformations by way of extensive exploration of the two and three dimensional cases;

2. Develop skill in computational aspects of linear algebra, especially with matrices and their uses;

3. Use concept and techniques of linear algebra to describe and analyze the geometry of R^2 and R^3;

4. Study applications of linear algebra in such areas as social science, business, and physical and biological sciences;

5. Learn some of the relationships of linear algebra to other areas of mathematics, such as linear programming, differential equations, and Markov chains.

While a linear algebra course may serve well as an introduction to more abstract mathematics, "the main goal is to emphasize applications and computational methods..." (CUPM Recommendations for a General Mathematical Sciences Program, p. 39). For prospective teachers it is especially important that the course have a strong flavor of geometrical interpretations and applications. The use of computers in studying numerical methods in linear algebra is strongly encouraged. The great diversity of applications is indicated in some of the references. It is essential that students see a variety of applications of topics covered in the course. To allow enough time to explore relationships between linear algebra and geometry, it may be necessary to treat other applications more briefly, but a few should be selected for presentation in depth.

The following list of topics is not intended to prescribe order; for example, as mentioned earlier, geometrical and other applications should permeate the course.

Topics

. MATRICES (6 hours)

Systems of equations and matrices, matrix algebra, row reduction, singular and nonsingular matrices, determinants, inverses.

. VECTOR SPACES (6 hours)

Definition and basic properties of finite dimensional vector spaces, R^2 and R^3 as vector spaces, subspaces and their geometrical interpretation, dependent and independent sets, bases and dimension.

. LINEAR TRANSFORMATIONS (6 hours)

Linear transformations and their matrix representations, kernel and image and their relation to subspaces, composition and inverses of transformations.

. INVARIANT SUBSPACES AND ORTHOGONALITY (5 hours)

Characteristic equations and roots, invariant subspaces, dot product, angles, perpendicularity, orthogonal bases, orthogonal transformations.

. AFFINE TRANSFORMATIONS (9 hours)

Affine transformations and properties they preserve, distance preserving mappings, rotations and reflections about the origin, matrix representations of affine mappings, uses in the study of similarity and congruence.

. APPLICATIONS (10 hours)

Examples drawn from such subjects as conic sections, differential equations, economics, game theory, genetics, linear programming, Markov chains, and population growth. (These should appear at appropriate places throughout the course.)

References

Anton, Howard. Elementary Linear Algebra (3rd edition). John Wiley and Sons, New York, 1981.

Birkhoff, G., and MacLane, S. A Survey of Modern Algebra (4th edition). Macmillan, New York, 1977. (Chapters 8, 9, 10)

Curtis, Charles W. Linear Algebra: An Introductory Approach (3rd edition). Allyn and Bacon, 1974.

Kolman, Bernard. Introductory Linear Algebra with Applications. Macmillan, New York, 1979.

Roberts, A. Wayne. Elementary Linear Algebra. The Benjamin/Cummings Publishing Co., Menlo Park, 1982.

Rorres, Chris, and Anton, Howard. Applications of Linear Algebra. John Wiley and Sons, New York, 1979.

Strang, Gil. Linear Algebra and Its Applications (2nd edition). Academic Press, New York, 1980.

Probability and Statistics

This course provides an overview of the basic theory and applications of probability and statistics. The emphasis is placed upon the understanding of only those theoretical foundations that can be effectively applied. Students desiring a more thorough introduction or planning for graduate courses in probability and statistics should take the year-long introductory sequence in probability and statistics. The prerequisites are Calculus 1 and an introductory computing course.

Objectives

The objectives of the course in Probability and Statistics are to provide teachers with the ability to:

1. Present data in visual form by use of tables and graphs;

2. Identify and discuss the most important measures of central tendency and dispersion;

3. Identify and describe some of the most important probability distributions;

4. Estimate various parameters, for example, means of populations, the difference of the means of two populations;

5. Test statistical hypotheses, in particular, hypotheses about means, proportions, the correlation coefficient of a bivariate population;

6. Design and analyze simple experiments and interpret the results;

7. Fit regression lines to data by the method of least squares;

8. Meaningfully use at least one statistical computer package.

Topics

While there may be considerable variation in the topics selected to achieve these objectives, the following seem to be particularly suitable to include in such a course:

. PROBABILITY (12 hours)

General probability (motivation, axioms and basic rules, independence), random variables, univariate density and probability functions, moments, Law of Large Numbers, standard distributions (normal, binomial, Poisson, exponential), Central Limit Theorem (without proof).

. ORGANIZING AND DESCRIBING DATA (6 hours)

Tables and graphs (frequency tables and histograms; bivariate frequency tables and the misleading effects of too much aggregation; standard line and bar graphs and their abuses; stem and leaf displays; box-and-whisker plots; spotting outliers in data), univariate descriptive statistics (mean, median and percentiles; variance and standard deviation; a few more robust statistics such as the trimmed mean), bivariate descriptive statistics (correlation; fitting lines--or, if computing resources permit, other curves--by least squares).

. DATA (6 hours)

Random sampling using a table of random digits, simple random samples, experience with sampling variability of sample proportions and means, stratified samples as a means of reducing variability; and experimental design. (Why experiment? - motivation for statistical design when field conditions for living subjects are present; the basic ideas of control and randomization (matching, blocking) to reduce variability).

. STATISTICAL INFERENCE (18 hours)

Statistics vs. probability (the idea of a sampling distribution; properties of a random sample, e.g., its mean is normally distributed for normal populations; the Central Limit Theorem); tests of significance (reasoning involved in alpha-level testing and use of p-values to assess evidence against a null hypothesis; one- and two-sample normal theory tests and (optional) chi-square tests for categorical data; comment on robustness, checking assumptions, and the role of design of experiment in justifying assumptions); point estimation methods (method of moments, maximum likelihood, least squares, unbiasedness and consistency); confidence intervals (importance of error estimate with point estimator; measure of size of effect in a test of significance); and inference in simple linear regression.

References

Bhattacharyya, G., and Johnson, R. Statistical Concepts and Methods. John Wiley and Sons, New York, 1980.

Box, G., Hunter, W., and Hunter, J. Statistics For Experimenters: An Introduction to Design, Data Analysis, and Model Building. John Wiley and Sons, New York, 1980.

Neter, J., Wasserman, W., and Whitmore, G. Applied Statistics. Allyn and Bacon, Boston, 1978.

Wonnacott, T., and Wonnacott, R. Introductory Statistics. John Wiley and Sons, New York, 1977.

Number Theory

This is a course covering some of the widely known theorems, conjectures, unsolved problems, and proofs of number theory. The prerequisite is two years of lower division mathematics of which one should have a strong algebra component (with emphasis on mathematical induction).

Objectives

The objectives of a number theory course should be to provide teachers with:

1. A better understanding and a deeper appreciation of some of the properties of the number system;

2. An opportunity to learn systematically the processes of problem solving and engage in extensive practice on this;

3. An opportunity to formulate and test conjectures;

4. An opportunity to develop, analyze, and practice theorem-proving skills in a concrete setting;

5. An opportunity to see concrete examples of abstract structures;

6. Acquaintance with some of the surprising and fascinating theorems of number theory which constitute one of the easiest and best devices for lending enrichment and excitement to their own teaching;

7. An appreciation of the usefulness of number theory in enhancing the teaching of school mathematics;

8. Familiarity with some of the most famous problems, solved or not, of number theory (these form such an important part of the history and tradition of mathematics) and with some of the key persons involved.

Topics

While there may be considerable variation in the topics selected to achieve these objectives, the following would seem to be particularly suitable to include in such a course:

. DIVISIBILITY (4 hours)

The division algorithm, the greatest common divisor, the Euclidean algorithm, the least common multiple.

. PRIMES (5 hours)

Unique factorization theorem, number systems without unique factorization, Euclid's proof of the infinity of primes, the distribution of primes, the prime number theorem (without proof), primes in arithmetic progressions, twin primes, prime triplets, the Goldbach conjecture.

. CONGRUENCES (10 hours)

Operations with congruences, proof of rules for divisibility by 3, 9, 11, etc. by use of congruences, the Euler ϕ-function, Fermat's theorem, Euler's generalization of Fermat's theorem, Wilson's theorem, solutions of congruences, the Chinese Remainder Theorem.

. DIOPHANTINE EQUATIONS (7 hours)

Linear Diophantine equations in two unknowns, positive solutions of such equations, linear Diophantine equations in several unknowns, the complete solution of $x^2 + y^2 = z^2$, sums of four squares, some simple

cases of Waring's problem, Fermat's last theorem.

. ARITHMETIC FUNCTIONS (8 hours)

The greatest integer function, the number of divisors of n, the sum

of positive divisors of n, the Moebius function.

In addition, it is suggested that one or two topics and related

applications from the following list be included:

. PARTITIONS

. FIBONACCI NUMBERS

. CONTINUED FRACTIONS

. REPRESENTATION BY SUMS OF TWO OR THREE SQUARES.

References

Davenport, H., The Higher Arithmetic (2nd edition). Cambridge University
Press, New York, 1982.

Dudley, Underwood, Elementary Number Theory (2nd edition). W. H. Freeman
and Company, San Francisco, 1978.

Long, Calvin T., Elementary Introduction to Number Theory (2nd edition).
D.C. Heath and Company, Lexington, 1972.

Niven, Ivan, and Zuckerman, Herbert S., An Introduction to the Theory of
Numbers (4th edition). John Wiley and Sons, New York, 1980.

Geometry

In this course Euclidean and other geometries are developed to provide one of the basic points of view for the study of mathematics. The prerequisite is the completion of two years of lower division mathematics.

Objectives

For prospective teachers the objectives of the geometry course should include the provision of sufficient experience in plane and three-dimensional geometry to enable them to:

1. Appreciate the use of geometry in various mathematical subjects;

2. Understand the relationships of geometry with other disciplines;

3. Present geometry as a source of mathematical models;

4. Present Euclidean geometry as a mathematical system and as one of several geometries;

5. Apply their knowledge of geometry as a mathematical model and as a mathematical system to

   (a) provide informal expository developments of the concepts of school geometry, and

   (b) attack mathematical problems and everyday problems from a geometric point of view.

The prospective teacher should be able to mention the occurrence of a wide variety of geometrical ideas outside of mathematics and especially in everyday life. These need not be extended accounts in depth but should at least be convincing and stimulating allusions to geometrical thinking in such areas as art, architecture, biology, cartography, crystallography, photography, physics, space exploration, sports, technology, and urban planning.

The content of the geometry course should complement the use of geometry in other courses in the prospective teacher's undergraduate program. For example, the use of linkages, geoboards, models of solids, and other manipulative devices should be explored in the methods course. Extensive use should be made of geometric approaches in the calculus courses, of geometric interpretations in the algebra courses, and of geometric representations in the modeling course. The history course should include the historical origins of geometry and the role of geometry in explorations of the physical universe, in the development of algebra, in the development of calculus, and in problem solving. The primary emphasis in the geometry course should be upon aspects of two- and three-dimensional Euclidean geometry that can be used in teaching high school geometry.

Topics

. INFORMAL APPROACHES (12 hours)

Explorations in, and informal approaches to, geometry are particularly important for prospective teachers. The use of such approaches in the geometry course provide prospective teachers with opportunities to learn ways of bridging the gap between the concrete models of geometry in the student's environment and the abstract geometric concepts and relationships discussed in class. The use of pencils for lines, pieces of cardboard for planes, computer graphics for solids and surfaces, and other manipulatives in explorations helps students relate the physical world to geometric concepts and strengthens their abilities in abstracting geometric relationships, checking the resulting patterns, and testing the validity of the generalizations formed. Some instructors may make extensive use of geoboards; others may prefer to emphasize paper folding and/or straightedge and compass constructions. Each approach should be included along with other favorites. Among the many desirable topics are:

--the illustration of numerous properties of geometric figures, for example using paper folding;

--observations of parallelism, perpendicularity, and distance in our three-dimensional physical universe; and

--elementary topological concepts such as betweenness, separation, continuity, neighborhoods on a plane and in space, limiting point and boundary.

Constructions of plane figures by paper folding should include the construction of points of a circle with a given diameter and the concurrence of the medians (altitudes, angle bisectors) of a triangle. Straightedge and compass constructions should include circles that intersect a fixed circle at right angles so that the model of circles orthogonal to a fixed circle may be used in the discussion of hyperbolic geometry.

. OTHER GEOMETRIES (4 hours)

The use of geometric representations provides a basis for informal discussions of spherical geometry, elliptic geometry, hyperbolic geometry, and a comparison of Euclidean, spherical, elliptic, and hyperbolic geometries.

. ALGEBRAIC METHODS (7 hours)

Coordinates in two or three dimensions should be used in the discussion of several topics and the proofs of appropriate theorems of high school geometry. Also vectors and transformations should each be used in discussions of relations among figures and the proofs of theorems.

. AXIOMATIC SYSTEMS (8 hours)

At most one third of the geometry course should be devoted to the study of high school geometry as a mathematical system. A synthetic axiomatic development of Euclidean geometry and the informal recognition of a hierarchy of geometries including topology, projective geometry, affine geometry, similarities, equiareal geometry, and Euclidean geometry provide important points of view on geometry for prospective teachers.

References

The recommended breadth of the scope of the geometry course makes it necessary to use more than one reference.

Adler, Claire Fisher, Modern Geometry (2nd edition). McGraw-Hill Book Company, New York, 1967.

Barry, E. H., Introduction to Geometric Transformations. Prindle, Weber and Schmidt, New York, 1966.

Coxeter, H.S. M., Introduction to Geometry (2nd edition). John Wiley and Sons, New York, 1969.

Eves, Howard, A Survey of Geometry (Revised edition). Allyn and Bacon, Boston, 1972.

Greenberg, Marvin Jay, Euclidean and Non-Euclidean Geometries, Development and History (2nd edition). W. H. Freeman and Company, San Franscisco, 1980.

Meserve, Bruce E., Fundamental Concepts of Geometry. Addison-Wesley Publishing Company, Reading, 1955; Dover Publications, New York, 1983.

Moise, Edwin E., Elementary Geometry from an Advanced Standpoint (2nd edition). Addison-Wesley Publishing Company, Reading, 1974.

National Council of Teachers of Mathematics, Geometry in the Mathematics Curriculum, Thirty-sixth yearbook NCTM, Reston, 1973.

Pedoe, Daniel, Geometry and the Liberal Arts. Penguin Books, Baltimore, 1976.

Smart, James R., Modern Geometries (2nd edition). Brooks/Cole, Monterey, 1978.

Yaglom, I.M., Geometric Transformations, translated from the Russian by Allen Shields, New Mathematical Library, vol. 8. The Mathematical Association of America, Washington, 1962.

_____, Geometric Transformations II, translated from the Russian by Allen Shields, New Mathematical Library, vol. 21. The Mathematical Association of America, Washington, 1968.

_____, Geometric Transformations III, translated from the Russian by Abe Shenitzer, New Mathematical Library, vol. 24. The Mathematical Association of America, Washington, 1973.

Young, G. C. and Young, W. H., Beginner's Book of Geometry. Chelsea, New York, 1970.

Abstract Algebra

A study of algebraic structures and their applications. The prerequisite is two years of lower division mathematics, including discrete mathematics.

Objectives

The objectives of this course are to provide teachers with:

1. Familiarity with algebraic properties of systems that appear in the high school curriculum: number systems, polynomials, mappings, symmetries, matrices;

2. Knowledge of definitions, examples and basic properties of certain algebraic structures: groups, rings, fields;

3. Experience in abstract reasoning, making and testing conjectures and proving theorems;

4. Awareness of applications of algebra in areas such as geometry, combinatorics, physics, mineralogy and computer science;

5. Acquaintance with the historical development of abstract algebra and of its relation to what is called high school algebra.

The objectives of this course may best be met by the Abstract Algebra II (modest) course described in the CUPM Recommendations for a General Mathematical Sciences Program. Such a course would be designed to "... show students at least two different types of algebraic structures and several instances in which such an algebraic structure evolved or is constructed out of another mathematical structure. The goal is for a student to be able to recognize when a situation has aspects that lend themselves to an algebraic formulation; e.g., rings out of polynomials" (p.48).

The course should include extensive work with examples, which can be used to make and test conjectures prior to the development of the formal theory. In proving theorems, an emphasis on the process of developing proofs and an analysis of proof techniques is important. Students are expected to be active participants rather than passive spectators of the development of the theory. It is not expected that the theory of all the algebraic structures introduced will be developed. The CUPM Recommendations suggest the following: "Describe part of the theory for one of the structures... and illustrate several of the deductive steps in the theory. Students should see the nature of tight logical reasoning and the usefulness of algebraic concepts, as well as come to appreciate the cleverness of the theory's discoveries" (p.48).

References to applications should occur regularly throughout the course and should be brought in as early as possible via examples. Models of molecules or crystals can be used in constructing some of the standard examples of groups. Reference to the many uses of matrices can be made when the topic of matrices as elements of a ring is introduced. The use of polynomials over a finite field in coding theory can be mentioned when integers modulo n are discussed.

Topics

The Panel believes that an algebra course taught in the spirit described above and covering topics of particular interest to prospective teachers could serve both teachers and other mathematics majors. The list of topics is therefore not a course outline; it merely specifies essential ingredients of an algebra course taken by prospective teachers. It is not intended to indicate order or interdependence of topics. Rings could just as well be listed prior to groups, for example. Similarly, equivalence classes could be introduced in the context of general set theoretic concepts or in conjunction with the definition of cosets. The specification of only 30 hours of a semester course allows considerable variation in emphasis. Additional time could be devoted to optional topics, to a more thorough development of portions of the theory, or to applications.

. GROUPS (9 hours)

Examples of groups and subgroups; permutation groups; groups of symmetries, especially those of quadrilaterals, regular polygons, the Platonic solids and the Rubik cube; cosets and Lagrange's Theorem.

. EQUIVALENCE RELATIONS AND MORPHISMS (4 hours)

Equivalence classes; integers modulo n; isomorphisms; homomorphisms and kernels.

. RINGS AND FIELDS (9 hours)

Integers, rational, real and complex number systems; the ring of matrices; polynomial rings, factoring polynomials, the Euclidean algorithm, finding zeros of polynomials; at least anecdotal reference to the history of solving polynomial equations, including efforts to solve the quintic equations and the related development of group theory.

. APPLICATIONS   (at least two 3 hour blocks)

Crystallography, crystal classification; algebraic coding theory, error correcting codes; applications of the Polya-Burnside Theorem; finite geometries.

. EXTENDING A GIVEN STRUCTURE TO A NEW STRUCTURE (2 hours)

Construction of factor groups; construction of a field from an integral domain; adjoining $\sqrt{2}$ or   ($\omega^3 = 1$ and $\omega \neq 1$) to Z.

In addition, the Panel suggests that at least one of the following topics be included:

. SEMIGROUPS AND APPLICATIONS

. FUNDAMENTAL HOMOMORPHISM AND ISOMORPHISM THEOREMS

. GALOIS THEORY

    Impossibility of certain Euclidean constructions, impossibility of a general solution of quintic equations.

References

Texts:

Durbin, John R., Modern Algebra, John Wiley and Sons, New York, 1979.

Gilbert, W.J., Modern Algebra with Applications, Wiley-Interscience, New York, 1976.

Lipson, John, Elements of Algebra and Algebraic Computing, Addison-Wesley, Reading, 1981.

Supplementary Sources:

Budden, F.J., The Fascination of Groups. Cambridge University Press, New York, 1972.

Burns, Gerald, Introduction to Group Theory with Applications, Academic Press, New York, 1977.

Cotton, F.A., Chemical Applications of Group Theory. Wiley-Interscience, New York, 1971.

Dornhoff, Lawrence and Hohn, Franz, Applied Modern Algebra. Macmillan, New York, 1978.

Nussabaum, Allen, Applied Group Theory for Chemists, Physicists and Engineers. Prentice-Hall, Englewood Cliffs, 1971.

Vincent, Alan, Molecular Symmetry and Group Theory, A Programmed Introduction to Chemical Applications. John Wiley and Sons, New York, 1977.

History of Mathematics

The history of mathematics is concerned with the origins, philosophy, and development of the mathematical sciences. The prerequisites are Calculus 1 and additional background acquired in a variety of other branches of mathematics.

Objectives

The objectives of the history of mathematics course are to provide teachers of mathematics with:

1. An understanding of mathematics both as a science and as an art (mathematics as a deductive science is emphasized in most mathematics courses; as an art, mathematics is a creative subject that includes the application of inductive insights and intellectual curiosity to the solution of problems and the formulation of theorems);

2. The ability to develop a broad concept of the mathematical sciences as approachable from several points of view including

    --problem solving, as a basis for the initial development of many concepts;

    --mathematics as a human endeavor, the role of individuals of both sexes with their insights and idiosyncrasies;

    --mathematics as a cultural heritage, the evolving role of mathematics in cultures throughout the world;

    --the impact of social, economic, and cultural forces on mathematical study and creativity;

--interrelations among the various branches of mathematics, especially their role in the solution of significant problems and in extending the horizons of mathematics; and

--the dynamic nature of mathematics, including the relatively recent development of probability and statistics and the increasing roles of calculators and computers;

3. Resources for developing the empirical and mathematical origins of each area of school mathematics including the notations, terminology, and major topics of algebra, geometry, trigonometry, calculus, number theory, probability, statistics, computer science and non-physical-science applications of mathematics. Such developments should be recognized as useful at all levels for organizing knowledge in historical perspective and appropriate in more detail for enrichment.

Topics

This history course should include, but not necessarily be structured according to, the chronological development of the mathematical sciences. The emphasis should be upon mathematical concepts and their interrelations. The suggested topics are a sample of the many available and worthwhile possibilities. The numbers of class hours are included to suggest approximate emphasis upon topics but are expected to vary according to the backgrounds and interests of the students.

. INFORMAL ORIGINS (4 hours)

Arithmetic and geometric concepts in early and primitive cultures, mathematical procedures based upon experiences and presented by examples,

early numeration systems, the nature and content of the Rhind
Mathematical Papyrus and Babylonian cuneiform tablets, extensive uses of
tables.

. THE EARLY DEVELOPMENT OF MATHEMATICS AS A SCIENCE (7 hours)

The work of the early Greek philosophers, the beginning of
demonstrative geometry, the Pythagoreans, figurate numbers, commensurable
and incommensurable magnitudes, Zeno's paradoxes, Eudoxus' method of
exhaustion and theory of proportion, the three famous problems, geometric
algebra, postulational thinking, Aristotle's laws and his concern for
foundations, Euclid's Elements.

. MATHEMATICS IN THE GREEK CULTURE AFTER EUCLID (6 hours)

Continued efforts to understand the physical universe stimulate many
aspects of the progress in mathematics. Note the continued development
of geometry, properties of numbers, an early form of algebra, and the
work of Archimedes, Eratosthenes, Apollonius, Hipparchus, Claudius
Ptolemy, Heron, Diophantus, and Pappus.

. MATHEMATICS OUTSIDE EUROPE BEFORE 1600 (5 hours)

The mathematics developed in Chinese, Hindu, Arabic and other
cultures. Suggested topics include rod numerals, the Chou-pei Suan
Ching, the algebraic methods of Chu Shih-chieh, the Sulvasutras, the
algebraic methods of Brahmagupta and Bhaskara, the algebra of
al-Khowarizmi and Khayyam, and the Arabic contributions through the
preservation, compilation, and extension of Greek and Hindu mathematics.

. MATHEMATICS IN EUROPE BEFORE 1600 (3 hours)

The introduction and transmission of Hindu-Arabic numerals,
translations of Euclid's Elements and other manuscripts, development of

trigonometry, development of arithmetic and algebraic notations, solutions of cubic and quartic equations, slow acceptance of decimal fractions, and extensions of synthetic geometry.

. MATHEMATICS IN THE 17TH AND 18TH CENTURIES (7 hours)

Analysis emerges first as algebra and then as calculus. Algebra acquires a significant role. Geometry gives way to analysis as the leading area of activity. Suggested topics are logarithms, theory of equations, theory of numbers, probability, statistics, analytic and synthetic geometry, the boost from science, calculus, infinite series, differential equations, and widespread explorations some of which, such as operations with divergent series, involved highly theoretical extensions.

. MATHEMATICS IN THE 19TH AND 20TH CENTURIES (7 hours)

Non-Euclidean geometry sets the stage for abstract mathematical systems that do not necessarily describe the physical universe. Geometry and algebra evolve into abstract mathematical systems. Set theory and logical foundations are developed. Gradually there is an increased recognition of the importance of each area of mathematics as a useful approach to the study of the mathematical sciences. Statistical methods, the use of computers, and a continued emphasis upon applications are important aspects of the ongoing development of the mathematical sciences.

References

Aaboe, Agar. Episodes from the Early History of Mathematics. New Mathematical Library, No. 13. The Mathematical Association of America, Washington, D.C. 1964.

Boyer, Carl B., A History of Mathematics. John Wiley and Sons, New York, 1968.

Cajori, Florian, A History of Mathematical Notations (2 vols.). Open Count Publishing Company, Chicago, 1928-29.

Eves, Howard, An Introduction to the History of Mathematics (5th edition). Saunders, Philadelphia, 1983.

_____, Great Moments in Mathematics (Before 1650). Dolciani Mathematical Expositions, Number Five, Mathematical Association of America, Washington, 1980.

_____, Great Moments in Mathematics (Since 1650). Dolciani Mathematical Expositions, Number Seven, Mathematical Association of America, Washington, 1982.

Hallerberg, Arthur E., et al. eds., Historical Topics for the Mathematics Classroom, Thirty-first Yearbook. National Council of Teachers of Mathematics (NCTM), Reston, 1969.

Kline, Morris, Mathematical Thought from Ancient to Modern Times. Oxford University Press, New York, 1972.

Menninger, Karl, Number Words and Number Systems: A Cultural History of Numbers. MIT Press, Cambridge, 1969.

Van der Waerden, B.L., Science Awakening, translated by Arnold Dresden. Oxford University Press, New York, 1961.

The books on the history of mathematics by Boyer and Eves are widely used as texts. The NCTM Yearbook and the book by Kline do not have exercises. The NCTM Yearbook contains seven articles and 120 capsules on historical topics that are of special interest to teachers.

Mathematical Modeling and Applications

This course is intended to provide the teacher with further experiences with the interaction of mathematical thinking with real-world problems and with a better understanding of the process as background for teaching applications of school mathematics. The prerequisites are two years of lower division mathematics, including discrete mathematics, calculus, and some computing. Additional preparation, for instance in probability and statistics, is desirable. Topics in the course should be chosen so as to give students opportunities to apply as much as possible of the mathematics they have learned.

Objectives

The objectives of the course should be to provide teachers with, or enhance:

1. Familiarity with a variety of applications of each of the major topics of school mathematics;

2. Sensitivity to opportunities to use mathematics in fresh ways to deal with real-world problems;

3. An acquaintance with and understanding of some of the best accounts of the nature of mathematical modeling, and the ability to analyze specific instances of it within the framework provided by those accounts;

4. Knowledge of reasonable standards for good textbook "story problems" and the ability to use those standards to evaluate problems and to create good new ones;

5. Familiarity with existing sources of applications of school mathematics;

6. A capacity for the infectious enjoyment of good applications;

7. Exposure to some mathematical ideas, for example in discrete mathematics, that may not be met in other courses;

8. Acquaintance with the principal mathematical needs of people in some of the trades, professions, and scientific disciplines.

Topics

The objectives just enumerated do not imply a specific syllabus. The choice of topics in any particular course may be guided by the interests of students and instructor, special regional concerns and opportunities, content of accessible instructional materials, etc. In making this choice the instructor should naturally strive for variety and balance as regards mathematical methods, areas of application, and levels of difficulty. In line with the objective 7, the approach of discrete mathematics should be given special consideration. A reasonable emphasis should be put on the use of calculators and computers.

In line with objectives 3, 4, and 5 especially, some time--perhaps six hours, preferably not at the beginning of the course--should be set aside for a survey of relevant literature, discussion of what might be called the philosophy of mathematical modeling, and similar activities.

A few of the many topics that might be included in such a course are listed below. Of course, not all of these topics need be covered in the course, and instructors are encouraged to find examples of their own.

. TAX INCENTIVES TO POLLUTION CONTROL

In recent years much legislation has been written to encourage private industry to install pollution control devices. The incentives are sometimes in the form of alternative patterns of taxes favorable to companies that have invested in pollution control. Interpretation and comparison of the alternatives offer interesting opportunities for the use of ideas like compound interest, depreciation, and (arithmetic and geometric) sequences.

. APPLICATIONS OF GRAPH THEORY TO ARCHAEOLOGY

The theory of graphs, especially interval graphs, may be used effectively to treat certain problems of seriation in archaeology. Very little mathematical background is required. (See D. G. Kendall, "Incidence matrices, interval graphs, and seriation in archaeology," Pacific Journal of Mathematics, 28, (1969), 565-570.)

. THE MEASUREMENT OF VOTING POWER

Rather simple combinatorial considerations lead to several plausible definitions of the power of an individual voter in a single- or multi-stage decision making system. These definitions and their consequences give interesting insights into the operation of such institutions as the U.S. electoral college and have been important in a number of recent court cases. (See, for example, Philip D. Straffin, Jr., Topics in the theory of voting, UMAP Expository Monograph Series, Birkhauser, Boston, 1980.)

. POPULATION

The size of populations over time offers many opportunities for modeling. There may be one or several species, competing or cooperating; the populations may consist of human beings, other large animals, insects, microorganisms, etc. The models may be deterministic or stochastic, and conclusions sometimes depend in interesting ways on whether variables are treated as discrete or continuous. The "predator-prey" models form an important special case which alone accounts for an enormous literature. This topic provides an excellent occasion for a careful analytic and geometric discussion of the "logistic" differential equation, which serves as a good example of the use of differential equations in applications.

. LINEAR PROGRAMMING

This has become a significant branch of mathematics in its own right, but may easily be treated as one topic among others. Mathematically, it usually deals with the problem of finding maxima and minima of linear functions under linear inequality constraints on the independent variables. It has applications in agriculture, management, transportation, military strategy, etc., any one of which might be a good starting point for a discussion.

. TRAFFIC LIGHTS

Red lights that last too long and lights at successive intersections that force frequent stops and starts are among the things that raise drivers' blood pressure and can retard traffic flow. Elementary mathematical considerations (and some not so elementary) can be used to time single traffic lights, or networks of lights, to minimize these effects; and allowances can be made for special lights for turns or for pedestrians.

. PACKING AND COVERING

Problems in this area include: cutting material for clothes or curtains; cutting pastry; buying, cutting, and placing floor coverings; designing containers for objects with a cylindrical, spherical, or other nonrectangular shape; arranging aisles, dividing lines, entrances and exits to turn a given piece of ground into a parking lot; designing refrigerators, closets, and other storage spaces; and so on. In each case one wishes to achieve a certain goal (such as covering a floor) without excessive waste.

This brief list emphasizes applications to areas other than the physical sciences and technology. These traditional applications should not be overlooked when the course is being planned.

Whatever topics are chosen, the course should require active participation from the students--as readers of the literature, as modelers, as critics, as fledgling teachers. How this will be accomplished will depend on the instructor's style, on the size of the class, and other factors. One technique that has been found stimulating and effective is to divide the class into small modeling teams, each of which is responsible for the development and presentation to the rest of the class of one or more models.

References

Published materials on which the course could be based are now plentiful, and there is a steady flow of new ones. An enthusiastic and qualified teacher of teachers should find planning the course with some of these materials challenging but feasible and rewarding.

The publications listed below contain many further references.

Aris, R., Mathematical Modelling Techniques, Research Notes in Mathematics 24. Pitman, London, 1978.

Bushaw, D., et al., A Sourcebook of Applications of School Mathematics. National Council of Teachers of Mathematics, Reston, 1980.

Haberman, Richard, Mathematical Models: Mechanical Vibrations, Population Dynamics, and Traffic Flow. Prentice-Hall, Englewood Cliffs, 1977.

Kaplan, Wilfred, "Broadening the scope of mathematics and its applications," Appl. Math. Notes 1, no. 2, 47-63 (1975).

Lancaster, Peter, _Mathematics: Model of the Real World._ Prentice-Hall, Englewood Cliffs, 1976.

Lucas, William F., et al. _Modules in Applied Mathematics_, 4 vols. Springer-Verlag, New York, 1983.

Noble, Ben, _Applications of Undergraduate Mathematics in Engineering._ The Mathematical Association of America, Washington, 1967.

Olinick, Michael, _An Introduction to Mathematical Models in the Social and Life Sciences._ Addison-Wesley, Reading, 1978.

Saaty, T. L., _Topics in Behavioral Mathematics._ The Mathematical Association of America, Washington, 1973.

Sharron, Sidney, ed., _Applications in School Mathematics_, 1979 Yearbook. National Council of Teachers of Mathematics, Reston, 1979.

The instructor of a modeling and applications course should certainly be aware of the publications of UMAP (Undergraduate Mathematics and its Applications Project); catalogs and other information may be obtained from EDC/UMAP, 55 Chapel Street, Newton, MA 02160. See also the quarterly _UMAP Journal_ (Birkhauser, Boston) as well as such other periodicals as the _International Journal of Mathematical Education in Science and Technology_ (Taylor and Francis, London) and _The Mathematics Teacher_ (National Council of Teachers of Mathematics, Reston).

In connection with objective 8, every mathematics teacher should be familiar with the leaflet "The Math in High School... You'll Need in College" (The Mathematical Association of America, Washington, n.d.).

Finally, teaching mathematical modeling to undergraduates is discussed at length in the CUPM booklet _Recommendations for a General Mathematical Sciences Program_ (The Mathematical Association of America, Washington, 1981), Chapter V. Further references are given there.